JOHN PHILLIP JAEGER

# SCIENCE
## *of the* BIBLE

JOHN PHILLIP JAEGER

# SCIENCE
## *of the* BIBLE

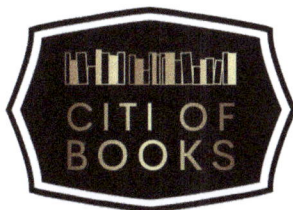

CITI OF
BOOKS

**CITIOFBOOKS, INC.**
3736 Eubank NE Suite A1
Albuquerque, NM 87111-3579
*www.citiofbooks.com*
Hotline:      1 (877) 389-2759
Fax:          1 (505) 930-7244

Ordering Information:

Quantity sales. Special discounts are available on quantity purchases by corporations, associations, and others. For details, contact the publisher at the address above.

Printed in the United States of America.

| ISBN-13: | Softcover | 979-8-89391-535-8 |
|---|---|---|
| | eBook | 979-8-89391-536-5 |
| | Hardback | 979-8-89391-534-1 |

# Contents

# Preface

As a chemical engineer, I understand and subscribe to the tenets of the scientific method. If anything has given me an appreciation for the profound fortuitous interdependencies* which make life possible, it is the objectivity engendered by my scientific background, coupled with common sense so often lacking in many well-educated people. How anti-intellectual it is of the atheist left to denigrate Christians, often maliciously so. Calling Christians *fundies* and *believers in a flat earth* seems to give many people the perverse notion that they are erudite and can consign Christians to the backwaters of ignorance. Just as Jesus overturned the tables of the money changers in his temple, so too is it my intention to overturn the tables of intolerant atheists with these personally written observations correlating science with its brilliant creator.

> Christianity is a fighting religion. (C. S. Lewis, author of *Mere Christianity*)[1]

> Do not think that I came to bring peace on earth. I did not come to bring peace but a sword. (Matthew 10:34)

> Then said he unto them, But now, he that hath a purse, let him take *it*, and likewise *his* scrip: and he that hath no sword, let him sell his garment, and buy one. (Luke 22:36)

*Atheist left* refers to the group which frequently and maliciously attacks the *religious right* as ignorant Bible-thumpers. For example,

---

[1] QuoteFancy

Googling *atheist left* produces only 56,400 hits versus 1,470,000 hits for *religious right*.

# CHAPTER 1

# The Beginning, Genesis 1:1

In the beginning God created the heaven
and the earth. (Genesis 1:1)

The Holy Bible was written more than two thousand years ago. In 1925, Edwin Hubble proved that the spiral nebula in the constellation Andromeda was a separate island universe, apart from the Milky Way[1]. This extended the size and scale of our universe by many orders of magnitude. Then, after hearing Albert Einstein's theory of relativity, Georges Lemaître, an ordained Catholic priest, proposed the *primeval atom* in 1927[2]—in other words, the creation of the universe. This breathtaking advancement in scientific thinking came not from a pontificating atheist, claiming to have exclusive jurisdiction over truth and science but rather from a devoted follower of Nature's God, as he is called in America's Declaration of Independence.

In 1929, Edwin Hubble discovered the redshift, eliminating any doubt that Lemaître was right and Einstein wrong. Einstein had said to Lemaître, "Your mathematics is correct, but your physics is abominable.[3]" This phenomenon, redshift, showed that some galaxies are moving away from us at greater speeds than others and that such velocities are proportional to their distance. This gave strong corroboration to the big bang theory of creation. The residual heat predicted in 1927 by Lemaître, and derisively dismissed by Albert Einstein, was later confirmed by Arno Penzias and Robert Wilson who in 1965 discovered the residual background radiation which is a remnant of the big bang.

---

[1] Physics of the Universe
[2] NASA.gov
[3] BigThinkBigBang

Penzias and Wilson[1] received the Nobel Prize for their discovery, which was accidental. The penning of Genesis 1:1 was not.

Prior to Lemaître's radical proposal, scientists believed that the universe was eternal and that it had always been as we see it today. An inherent aspect of the steady-state universe was the assumption that matter is continuously being created, somewhere, somehow. This passed for science, until it was disproved in 1965.

So we see twentieth century confirmation of the deep science originally expressed in the first sentence of the first paragraph of the first book of the Bible and scientifically advanced centuries later by a Catholic priest before anyone else.

> There is no conflict between religion and science. (Georges Lemaître[2]; coherence, correspondence, correlation)

> And the Lord God formed man of the dust of the ground and breathed into his nostrils the breath of life; and man became a living soul. (Genesis 2:7)

> In the sweat of thy face shalt thou eat bread, till thou return unto the ground; for out of it wast thou taken: for dust thou art, and unto dust shalt thou return. (Genesis 3:19)

Modern chemistry could not have begun before 1802, when John Dalton formally provided experimental evidence that matter is composed of discrete atoms.[3] Before this, it was all guesswork. Nevertheless, it was clearly stated in Genesis that man is "formed of the dust of the ground," which is to say, the same elements of carbon, oxygen, hydrogen, iron, nitrogen, etc. that we find in dust of the ground, minerals.

---

[1] Penzias
[2] Gizmodo
[3] Science History

And out of the ground the Lord God
formed every beast of the field, and every foul
of the air. (Genesis 2:19)

The same elements which form humans also formed animals everywhere. However, there was no biblical reference to *a living soul* with respect to animals. Nor do animals have the capacity to worship which is the premier hallmark of mankind and our supreme bequest.

Spirituality has been scientifically shown to contribute to mental and physical health, well-being, and longevity, and not only in humans. Experiments with Norwegian wharf rats showed that they drown in sixteen minutes, on average, when put in a bucket of water. But after being rescued at the last minute, dried, fed, and rested (given hope), the next day control rats would swim for sixty hours.[1] Hope operates in animals and humans. Hundreds of thousands of prisoners have been released from prison with newfound spiritual hope. I have never heard of a prisoner being paroled and exclaiming, "I found atheism in prison, and now I have hope for a new life!"

The earth also was corrupt before God, and
the earth was filled with violence. (Genesis 6:11)

One would think that as a result of the disciplines and analyses and benefits of human enlightenment, mankind should have been able to eliminate corruption and violence so prevalent thousands of years ago. Today, we have tools of production and health and social enlightenment unimaginable when the book of Genesis was written. But the earth today is still full of corruption and violence. Cornucopias of goods and services have not satisfied mankind's lust for more nor have psychologists and sociologists resolved the complex issues that lead people into destructive behavior. With burgeoning prison populations and monstrous acts of evil on the increase worldwide, there seems little hope that corruption, and violence will ever be eradicated.

---

[1] World of Work

> And the waters prevailed exceedingly upon
> the earth; and all the high hills, that were under
> the whole heaven, were covered. (Genesis 7:19)

Although the North American continent was unknown when the Bible was written, paleontologists confirmed that the interior of North America was once covered by shallow seas. Fossil evidence from distant parts of the globe that were unknown to inhabitants of ancient Israel lent scientific confirmation to the Noachian flood described in the ancient book of science, the Holy Bible. I do not pretend to know the length of the six *days* of creation.

However, it is abundantly clear to me that the elegance of everything, which I call *profound fortuitous interdependencies* in my book, *Brilliant Creations: The Wonder of Nature and Life*, combined with the insuperable statistics of naturalistic protein synthesis and the anthropic principle are eternally inexplicable by any means other than our Creator. To those with true discernment, God's hand is clearly visible everywhere one looks. This is evidenced by the fact that over 65 percent of Nobel Laureates were Christians, and another 20 percent were Jews, while under 11 percent were atheists or agnostics (*100 Years of Nobel Prizes*, Baruch Shalev, 2005, Los Angeles).

# CHAPTER 2

# Who Made God?

> But now the Lord saith, 'Be it far from Me;
> for them that honor Me I will honor, and they
> that despise Me shall be lightly esteemed. (1
> Samuel 2:30)

Honor is a deeply virtuous term, unlike terms so often used by deniers of Nature's God. Terms atheists commonly abuse include *intelligence, rational, evidence,* and *proof.* Meditate on these things of good report, as stated in Philippians 4:8, things that are true, noble, just, pure, and of course honorable.

The psychological and physical benefits of positive thoughts have been thoroughly documented by modern science[1] and are the subject of countless books. This is one such.

> Let your light shine before men in such a
> way that they may see your good works and
> glorify your Father who is in heaven. (Matthew
> 5:16)

> And as ye would that men should do to you,
> do ye also to them likewise. (Luke 6:31)

> Give, and it shall be given unto you; good
> measure, pressed down, and shaken together,
> and running over, shall men give into your
> bosom. For with the same measure that ye mete

---

[1] Mayo Clinic

withal it shall be measured to you again. (Luke 6:38)

Let your moderation (kindness, generosity) be known to all men. (Philippians 4:5)

Finally, brethren, whatever things are true, whatever things *are* noble, whatever things *are* just, whatever things *are* pure, whatever things *are* lovely, whatever things *are* of good report, if *there is* any virtue and if *there is* anything praiseworthy—meditate on these things. (Philippians 4:8)

And let us not be weary in well doing. (Galatians 6:9)

And let us consider one another to provoke unto love and to good works. (Hebrews 10:24)

Think positive. Be optimistic. Give your best and expect the best. It's biblical and it has valid scientific basis through research. God has forgiven you; you must forgive yourself. Put away and reject negative thoughts and guilt. It's hard; be persistent.

The greatest discovery of any generation is that human beings can alter their lives by altering the attitudes of their minds. (Albert Schweitzer, MD, Nobel Laureate)[1]

We scientists have found that doing a kindness produces the single most reliable momentary increase in well-being of any exercise we have tested. (*Flourish*, by Martin Seligman, Psychologist)[2]

---

[1] Goodreads
[2] Parentotheca

If you want to be happy, do good, be kind, give. Professor Arthur C. Brooks conducted research into generosity and its effects and wrote *Who Really Cares: America's Charity Divide: Who Gives, Who Doesn't, and Why It Really Matters.* He found that people of faith donate considerably more than do atheists, not only to churches but also to family, friends, and secular organizations. Believers also donate more of their volunteer time and even their blood. It not only makes them feel better but truly makes them better.

> Psychological professionals in the twenty-first century have affirmed the substantial benefits, both mental and physical, of positive thinking in countless scholarly experiments, published papers and books. (*Optimism and Its Impact on Mental and Physical Well-Being* [nih. gov])

> I am hath sent me unto you. (Exodus 3:14)

> Jesus said unto them, Verily, verily I say unto you, Before Abraham was, I am. (John 9:58)

The scientific approach to explaining how man and energy and matter and space originated is to examine what is observable and formulate hypotheses and theories based on observation and reason. There is, and can never be, any explanation for the origin of matter, energy, and information at the moment of creation which excludes the Creator.

Obviously, no experiment can examine, much less confirm any hypothesis of what first happened to lead to us and everything we see.

God defies scientific explanation because he is outside its domain. After all, God created the physical realm that is the subject of scientific inquiry, and we are still desperately trying to understand that limited aspect of His handiwork. Had mere mortals written where God came from without divine inspiration, they surely could not have presented such an elegant explanation as *I am*—an explanation that suffices even two thousand years later. Where did God come from? *I am*. The universe is not eternal, but God is. So are we.

> "But who made God?" chortle atheists, as if believers have no answer. "If anyone made God, then He wouldn't be God, would He?" (Mathematics Professor John Lennox, of Oxford University)[1]

Oxford's motto is, "The Lord is My Light."[2]

I have only a vague notion of how my computer works as I type this on it. Although I don't know how it works, I do know that it operates in a marvelous way. I don't need to understand things to believe in and use them. And how much more marvelous is my brain and yours than these primitive computers, not one of which designed, much less built itself.

> Many people don't realize that science basically involves assumptions and faith. Wonderful things in both science and religion come from our efforts based on observations, thoughtful assumptions, faith and logic. (With the findings of modern physics, it) seems extremely unlikely (that the existence of life and humanity are) just accidental. (Charles Townes, Nobel Laureate and Professor of Physics at University of California, Berkeley)[3]

> Because that which may be known of God is manifest in them (people); for God hath shewed it unto them.
> For the invisible things of him from the creation of the world are clearly seen, being understood by the things that are made, even his eternal power and Godhead; so that they

[1] Lennox
[2] Uni of Oxford
[3] NBC

(nonbelievers) are without excuse. (Romans 1:19–20)

"Show me proof of God," nonchalantly demanded by atheists, was anticipated and answered two thousand years ago. "The invisible things of him are clearly seen."

I spoke with a physics teacher about these observations, and he replied, "I ask my students, 'How long do you think it would take a warehouse of Tesla car parts to assemble themselves?'"

As creative and illustrative as this Socratic method question is, the far bigger challenge is for the warehouse full of complementary and perfectly matched parts to mine their ores and materials from around the world; ship themselves across oceans to factories where they would have to refine, shape, machine, and fabricate themselves to incredibly close tolerances; then ship themselves again to the final warehouse where they would then, hypothetically, assemble themselves.

> Through my scientific work I have come to believe more and more strongly that the physical universe is put together with an ingenuity so astonishing that I cannot accept it as a brute fact... I cannot believe that our existence in this universe is a mere quirk of fate, an accident of history, an incidental blip in the great cosmic drama. (Paul Davies)[1]

> When I began my career as a cosmologist some twenty years ago, I was a convinced atheist. I never in my wildest dreams imagined that one day I would be writing a book purporting to show that the central claims of Judeo-Christian theology are in fact true, that these claims are straightforward deductions of the laws of physics as we now understand them. I have been forced into these conclusions by the

---

[1]     AZ Quotes

inexorable logic of my own special branch of physics. (Frank J. Tipler)[1]

Some of leading scientists whose work was motivated by their faith were Copernicus, Kepler, Galileo, Brahe, Descartes, Boyle, Newton, Leibniz, Gassendi, Pascal, Mersenne, Cuvier, Harvey, Dalton, Faraday, Herschel, Joule, Lyell, Lavoisier, Priestley, Kelvin, Ohm, Ampere, Steno, Pasteur, Maxwell, Planck, and Mendel.

> God understandeth the way thereof, and he knoweth the place thereof. For he looketh to the ends of the earth, and seeth under the whole heaven; To make the weight for the winds; and he weigheth the waters by mea- sure. (Job 28:23–25)

Ask the average person what air weighs, and he will answer, "Nothing." However, the weight of a cubic mile of air at sea level is over five million tons.

> He stretcheth out the north over the empty place, and hangeth the earth upon nothing. (Job 26:7)

Has anyone the slightest doubt as to how *empty* the North Pole is? Nobody living in the Middle East could possibly have visited *the north*, so as to confirm what was then being written. These immutable scientific truths—two here in a single sentence—were far too coincidental to be attributable to luck. No, they were divinely inspired, as were so many other messages in the Bible. The earth truly hangs *upon nothing*, as is common knowledge today, shown in countless photographs from satellites and space stations, not to mention men on the moon. The north is indeed an *empty place* by any measure.

---

[1]      Tipler

> Lo, these are parts of his ways: but how little a portion is heard of him? But the thunder of his power who can understand? (Job 26:14)

> Then I beheld all the work of God, that a man cannot find out the work that is done under the sun: because though a man labour to seek it out, yet he shall not find it; yea further; though a wise man think to know it, yet shell he not be able to find it. (Ecclesiastes 8:17)

With all our wisdom, all our science, and all our research, *who can understand* anything completely today? Ultimate scientific answers continue to elude us everywhere one looks, from the submicroscopic to the super-macroscopic. The atheist's pretense is that all this magnificent science that we see and study arose from nothing, based solely on mega-time and multi-universes. Insuperable statistical impossibilities are explained away with clever wordplay and nebulous theories—anything at all to deny the hand of the Creator so evident to other observers, of all educational backgrounds, all nationalities, and all times, that is, except for those who believe clever words they have heard, that cannot stand up to thoughtful scrutiny.

> The larger the island of knowledge, the longer the shoreline of wonder. (Ralph W. Sockman)[1]

> The first gulp from the glass of natural sciences will turn you into an atheist, but at the bottom of the glass, God awaits you. (Werner Heisenberg, Nobel Laureate,[2] 1932)

> As for the earth, out of it cometh bread: and under it is turned up as it were fire. (Job 28:5)

---

[1] EagerEyes
[2] Goodreads Heisenberg

The molten iron core of the earth was inconceivable because it was not discoverable when this passage was written. *Under (earth) it is turned up as it were fire.*

Ah, some may say, "But there were volcanoes even then." True enough. But are not volcanoes both isolated and rare and not so much *under* the earth as above it? The molten core of the earth accords far better with this passage. Their scientific agreement is not coincidental but rather divinely inspired and guided.

> And unto man he said, Behold, the fear of
> the Lord, that is wisdom; and to depart from
> evil is understanding. (Job 28:28)

Knowledge is not wisdom nor is a high IQ. The Unabomber was a genius. Countless men and women throughout history have been filled with knowledge, while doing foolish and evil things which destroyed the lives of hundreds of millions of others.

> Then the Lord answered Job out of the
> whirlwind, and said, "Who is this that darkeneth
> counsel by words without knowledge?" (Job
> 38:1–2)

How often one hears words uttered without knowledge by pretenders of science and enlightenment.

> "I think it is virtually certain that
> everything we see came from empty space,"
> Krauss exposited. "And all the physics I know is
> highly suggestive that our universe popped into
> existence as a quantum fluctuation." (Physicist
> Lawrence Krauss)[1]

The Lord does not take foolishness lightly, and neither should we.

---

[1] Reasons

# CHAPTER 3

# Where Were You?

> Where were you when I laid the foundations
> of the earth? Tell *Me*, if you have understanding.
> Who fixed its measurements? (Job 38:4–5)

What *foundations of the earth* you may wonder? The very structure of matter, which is so elegantly created that water, glass, wood, iron, rocks, and your very body, is 99.9999999 percent empty space, between atomic nuclei and the quantized fluctuating probability wave functions, commonly called *electrons*.[1]

Who fixed the *measurements* of physical constants, from the gravitational constant to the electromagnetic constant, to the strong and weak nuclear constants? Who fixed the elegant slow speed of sound and enormously high speed of light? Why is the vast interconnectedness of nature and forces and processes so elegant and beneficial to humans? They were made that way by our brilliant Creator or Nature's God, as he is named in America's Declaration of Independence. Profound fortuitous interdependencies surround you if you will only pause to consider them and appreciate their Creator.

> Them that honor me I will honor. (1 Samuel 2:30)

> By what way is the light parted, which scattereth the east wind upon the earth? (Job 38:24)

---
[1] WTAMU

The most learned scientist of antiquity could not have imagined the depth of this question. When light is *parted* by a diffraction grating, it can be shown to act both as particles as well as waves. These combinations of properties are difficult to explain, much less really understand. The prodigious amounts of energy transmitted by solar radiation do indeed scatter the wind upon the earth as it heats different substances at different rates. Job could not have offered an adequate answer to the question, along the lines of: "Discrete photons of light travel together as a wave until parted into disparate visible components by striking and reflecting from solid objects into our eyes, while other wavelengths give up their energy as they are absorbed by solids and water. Temperature differentials established by ambient sunlight striking dissimilar surfaces create 'the east wind' so described."

> Can you send forth lightnings that they may go and say to you, "Here we are?" (Job 38:35)

In November 1886, Heinrich Hertz became the first person to transmit and receive controlled radio waves,[1] *sending forth lightnings* that say to you, "Here we are." Billions around the world now take for granted telephone transmissions through the air. This was inconceivable when the Bible was written.

> The heavens declare the glory of God; and the firmament sheweth his handywork. (Psalms 19:1)

NASA maintains a website which is updated daily. Even if unintended, it does indeed *declare the glory of God* and *show his handiwork*: http://antwrp.gsfc.nasa.gov/apod/astropix.html.

How remarkable that so repetitive and well-known phenomenon as sunset can delight people of all ages, all times, and all civilizations. How much more delightful are the glories and handiworks seen in

---

[1] Famous Scientists

national parks and sightseeing attractions worldwide, so many of which could scarcely have been known by the Bible's authors.

The earliest record of a telescope was in 1608.[1] How is it that the more deeply we have seen, the more handiwork has appeared? It could only be the marvelous design of the brilliant Creator. These wonders surrounding us daily are not trivial nor are they few in number.

Day unto day uttereth speech, and night
unto night sheweth knowledge. (Psalms 19:2)

For thousands of years, astronomers have studied planets and stars, which have spoken eloquent volumes of knowledge and even mathematics. Sir Isaac Newton invented the calculus in order to understand elliptical orbits of planets. No less a popular scientist than Carl Sagan said, "Astronomical spectroscopy is an almost magical technique. It amazes me still" (*Cosmos*, page 93). Whenever a person of faith expresses amazement, they are ridiculed for expressing the *argument from incredulity*.[2] But should any atheist or agnostic such as Sagan expresses the very same *argument*, other nonbelievers never ridicule or mock it. Such hypocrisy is unscientific and unintelligent.

---

[1]   Space
[2]   Logically Fallacious

## CHAPTER 4

# Rejoice and Be Glad

This is the day the Lord hath made; We will rejoice and be glad in it. (Psalms 118:24)

"Then I commended mirth, because a man hath no better thing under the sun, than to eat, and to drink, and to be merry: for that shall abide with him of his labour the days of his life, which God giveth him under the sun." (Exodus 8:15)

I know that nothing is better for people than to be happy, and to do good while they live. (Ecclesiastes 3:12)

Do not be anxious about anything, but in every situation, by prayer and petition, with thanksgiving, present your requests to God. (Philippians 4:6)

Take therefore no thought for the morrow: for the morrow shall take thought for the things of itself. Sufficient unto the day *is* the evil thereof. (Matthew 3:4)

Be strong and of a good courage; be not afraid, neither be thou dismayed: for the Lord thy God is with thee whithersoever thou goest. (Joshua 1:9)

Twenty-first-century medicine confirms the benefits of joyfulness for both our mental health and our physical health.

Pessimistic people tend to view problems as internal, unchangeable, and pervasive, whereas optimistic people are not so cynical. Pessimism has been linked with depression, stress, and anxiety (Kamen and Seligman 1987), whereas optimism has been shown to serve as a protective factor against depression, as well as a number of serious medical problems, including coronary heart disease (Tindle et al. 2009). Optimistic mothers even deliver healthier, heavier babies (Lobel et al. 2000). Optimism seems to have a tremendous number of benefits; consider several which are detailed below.

## Optimism and Physical Health

Few outcomes are more important than staying alive, and optimism is linked to life longevity. Maruta et al. (2000) examined whether explanatory styles served as risk factors for early death. With a large longitudinal sample collected in the mid-1960s, the researchers categorized medical patients as optimistic, mixed, or pessimistic. Optimism was operationalized using parts of the Minnesota Multiphasic Personality Inventory. The researchers found that for every 10-point increase in a person's score on their optimism scale, the risk of early death decreased by 19 percent. Considering that, for a middle-aged person of average health, the difference between sudden death risk factors for smokers and nonsmokers is only 5 to 10 percent, the protective effect of optimism found in this study is considerable.

Optimism also plays a role in the recovery from illness and disease. Multiple studies have investigated the role of optimism in people undergoing treatment for cancer (e.g., Carver et al. 1993; Schou et al. 2005). These studies have found that optimistic people experience less distress when faced with potentially life-threatening cancer diagnoses. For example, Schou and colleagues (2005) found that a superior *fighting spirit* found in optimists predicted substantially better quality of life one year after breast cancer surgery.

Optimism also predicted less disruption of normal life, distress, and fatigue in one study of women who were undergoing painful treatment for breast cancer (Carver et al. 2003). In this case, optimism appeared to protect against an urge to withdraw from social activities, which may be important for healing. People who tend to be more optimistic and more mindful had an increase in sleep quality (Howell et al. 2008). There is also evidence that optimism can protect against the development of chronic diseases. A sample of middle-aged women was tested for precursors to atherosclerosis at a baseline and three years later. The women who endorsed greater levels of pessimism at the baseline assessment were significantly more likely to experience thickening arteries while optimistic women experienced no such increase in thickness (Matthews et al. 2004) (pursuit-of-happiness.org).

> I will praise thee; for I am fearfully and wonderfully made: marvelous are thy works; and that my soul knoweth right well. (Psalms 139:14)

Mankind is indeed *fearfully and wonderfully made*. The sophistication of our construction begins at the atomic level with atoms that are one part in $10^{15}$ [1] nucleus and the rest empty space (*$10^{15}$* is an exponential notation, which is shorthand for *1* followed by fifteen zeroes). Then consider our DNA, which is forty-five trillion times more compact and efficient at data storage than today's sophisticated computer microchips.[2]

Our brains have the memory capacity of one hundred billion megabytes,[3] which far exceeds anything conceivably necessary from an evolutionary *selection* point of view.

Our optic nerves transmit information from our eyes to our brains at approximately ten million bits per second.[4]

The human eye sees in exquisite detail over about thirteen orders of magnitude of light intensity.[5] Although the eye is often said to be flawed in its design by Darwinists and atheists, I would very much like

---

[1] BigThinkEmpty
[2] Upper Cervical Institute
[3] AIImpacts
[4] Science Daily
[5] Neuron Research

to see them replace a human eye with something better which they have designed and built from lab reagents.

Our ears hear over twelve orders of magnitude in sound intensity.[1] Even more amazing, the ears perform a Fourier analysis.[2] In other words, our eardrums receive a single wave function at the eardrum. Then they break down this single wave function into its constituent sounds. For example, at a concert, your brain hears drums, brass, violins, solo arias, and the person behind you coughing, only because this combined sound is separated inside your ear. If your eyes performed a similar function, they would break down white, just as a prism does, into disparate pure colors.

Our eyes enable us to discern distance (and relative size) by triangulation. Our brains automatically compute the angle of the object seen and compute its approximate distance. Similarly, our two ears enable us to discern the direction from which noises emanate not only because we have two ears but also because of the relatively slow speed of sound. A difference in the arrival time from one ear to another of one thousandth to one ten thousandth of a second is sufficient to discern so that we can tell generally where a sound originated. If sounds were substantially faster, both our ears would hear the sound at about the same time, and we could not enjoy stereophonic music nor tell where sounds came from.

> Nature is God's greatest evangelist. (Jonathan Edwards [1703–1758], theologian and philosopher)[3]

In his wisdom, God made these velocities extremely useful to us, as well, of course, as many other physical constants. They did not *evolve* to such values.

Biochemistry is extremely complex, and we now understand more deeply than ever before how *wonderfully* we are made. Human blood defies Le Chatelier's principle in that when one molecule of oxygen is adsorbed by a hemoglobin molecule, its affinity for oxygen increases

---

[1]    Web.MIT
[2]    Brittanica
[3]    Love Expands

instead of diminishes. The second molecule increases the affinity for the third and the third for the fourth. This is precisely the reverse of normal chemistry principles and experimental observations. Our bodies' powers of endurance and healing are absolutely astounding.

The list of features of our wonderful construction begins with conception, continues through growth, and concludes with our spiritual transformations which have been evidenced time and again by the scientific observations of such people as Elisabeth Kübler-Ross. Dr. Kübler-Ross documented hundreds of instances of scientific evidence of human spiritual nature. She convincingly testified that she could not be persuaded of any naturalistic (scientific) explanation for it, such as hallucinations, for example.

> Happy is the man that findeth wisdom, and the man that getteth understanding. (Proverbs 3:13)

Twenty-first-century social science confirms that Christians are happier than atheists. When confronted with this scientific truth, atheists angrily and snidely replied, "Ignorance is bliss." It's a lie, and lies are all they have. To contend otherwise is to claim that wisdom decreases your happiness. Why then do we strive for and achieve wisdom?

# CHAPTER 5

# Get Wisdom

> By his knowledge the depths are broken up,
> and the clouds drop down the dew. (Proverbs
> 3:20)

Not only does the mid-Atlantic ridge constitute the continuing breakup of tectonic plates but also in the depths of the Pacific Ocean *the depths are broken up*, as discovered by modern science, unknown almost two thousand years ago.

> My son, let them not depart from your eyes.
> Keep sound wisdom and discretion. (Proverbs
> 3:21)

> These six things doth the Lord hate: A false
> witness that speaketh lies, and he that soweth
> discord among brethren. (Proverbs 6:16, 19)

> How much better is it to get wisdom than
> gold! And to get understanding rather to be
> chosen than silver. (Proverbs 16:16)

There are repeated biblical references to wisdom, prudence, diligence, and other intellectual virtues. The essence of science is the pursuit of truth, knowledge, and facts. However, wisdom requires more than mere knowledge. Wisdom requires the integration of science with integrity and far more virtuous conduct than the mere gathering of facts, which are value neutral. The very word *science* is derived from *Scientia*, Latin for *knowledge*.[1]

The Puritans were devout Christians who in 1635 established Boston Latin School, the first public school in the New World. The following year, Puritan clergyman John Harvard founded Harvard College, America's most prestigious university today. In 1650, the Harvard Corporation chose the college motto, "Christi Gloriam," Latin, meaning "For the glory of Christ."[2] That motto remained until 1836 when *veritas* (*truth*) reemerged from old documents.

Every Ivy League college has a Christian charter—a fact assiduously ignored by atheists feigning virtuous intellectualism.

Hitler and Stalin appreciated the pseudoscience of Darwinism for its atheistic implications, but of wisdom, they had none.

> *Atheism is the natural and inseparable part of Communism.* (Vladimir Lenin)[3]

> *I wish to avenge myself against the One who rules above.* (Karl Marx)[4]

> Forsake the foolish and live; and go in the way of understanding. (Proverbs 9:6)

> He that reproves a scorner gets himself to shame. (Proverbs 9:7)

> Go from the presence of a foolish man. (Proverbs 14:7)

You are known by the company you keep. The evil and foolish will drag you down to their level. Scientific experiments have confirmed the tendency of people to reject what they plainly know to be true in order to conform to groupthink (Janis 1972).

---

[1] Etymonline
[2] The Crimson
[3] QuotesLyfe
[4] Way of Life

> Train up a child in the way he should go;
> even when he is old he will not depart from it.
> (Proverbs 22:6)

Good parenting is critical, as is good teaching. Impressions that are ingrained in us tend to persist.

Confirmation bias is a well-established phenomenon for all people of all education levels. It is so difficult to overcome that Nobel laureate and physicist Max Planck said, "A new scientific truth does not triumph by convincing its opponents and making them see the light, but rather because its opponents eventually die, and a new generation grows up that is familiar with it."[1]

> Give me four years to teach the children
> and the seed I have sown will never be uprooted.
> (Vladimir Lenin)[2]

> And I gave my heart to seek and search out
> by wisdom concerning all things that are done
> under heaven. (Ecclesiastes 1:13)

> So teach us to number our days, that we
> may apply our hearts unto wisdom. (Psalms
> 90:12)

*To seek out and search by wisdom.* Isn't this what scientists do?

Wisdom is defined as the ability to use your knowledge and experience to make good decisions and judgments. Did Bernie Madoff, Harvey Weinstein, Jeffrey Epstein, Phil Spector, and so many other rich, famous, and well-educated people apply their hearts unto wisdom? No. Instead, they demonstrated the truth of "What does it profit a man to gain the whole world and lose his own soul" (Mark 8:36).

Knowledge may or may not be the truth. Scientists have given the world much knowledge that was untrue. The common claim that *science has a self-correcting mechanism*, so often smugly parroted by atheists, is

[1] (Roughly) Daily
[2] NewsWithViews

in fact trivial. All living organisms have this *self-correcting mechanism*. It is trial and error. If the search for food, water, or shelter in one direction fails, every organism performs the *self-correcting mechanism*. It changes direction and searches elsewhere.

It is incumbent on us to use our wisdom to separate the wheat from the chaff of cleverly worded plausible opinions.

The origin of plausible is *plausibilis*, Latin for "worthy of applause."[1] You may smile and applaud the lies of deceivers, but that does not make them so.

---

[1] Merriam-Webster

CHAPTER 6

# Water, Wow!

> All rivers run into the sea; yet the sea is not
> full; unto the place from whence rivers come,
> thither they return again. (Ecclesiastes 1:7)

The cyclical nature of nature encompasses us wherever we look. The water cycle is described only with utmost brevity in Ecclesiastes. Today we understand (considerably better than did biblical authors) water and its importance in most chemical processes and also its extremely beneficial cyclical nature, as only hinted in the Bible. Beyond this, we can see and describe cycles of carbon, nitrogen, oxygen, and hydrogen. We are able to understand the nature of energy in its many forms sufficiently to use its many forms for our sustenance and well-being. One form is our continuous sunlight.

Why should all these things be? And why so reliably? Why are chemical reactions so wonderfully and perfectly reversible? For the same reason that we are *fearfully and wonderfully made*. These Profound Fortuitous Interdependencies did not just fall into place with mega-luck over mega-time, as a small minority of scientists claim.

> Then I saw that wisdom excelleth folly,
> as far as light excelleth darkness. (Ecclesiastes
> 2:13)

> He hath made everything beautiful in his
> time: also he hath set the world in their heart,
> so that no man can find out the work that

God maketh from the beginning to the end.
(Ecclesiastes 3:11)

Can you name even one aspect of our universe that is known *from the beginning to the end*? Research continues everywhere, with absolutely no end remotely in sight.[1] Research at the subatomic level, at the molecular level, at the cellular level, at the human level, at the planetary level, and at the galactic level proceeds worldwide.

How is it that we don't know everything about anything if all that we see arose from nothing, as materialists contend? How is that possible? Infinite complexity from nothing is infinitely absurd and infinitely improbable. There is not the slightest scientific basis or law for their grand proposal. There is zero empirical evidence (meaning observed somewhere) of nothing making wonderful systems and animals.

Organization, information, consistency, and precise physical laws, are just a few elements that show how an infinite God makes scientific as well as spiritual sense. He is utterly beyond the purview of science, which is merely another of God's brilliant creations.

Anger rests in the bosom of fools.
(Ecclesiastes 7:9)

Let us not be desirous of vainglory, provoking one another, envying one another.
(Galatians 6:26)

Certainly not all, but many atheists are angry and vulgar. Atheism is an essential component of socialism and communism. The vainglory of atheists is prominent in their claims of intellectual and moral superiority. They cynically claim that Christians only do good to get to heaven, while the atheists themselves are good just because of their sanctimonious egos. Is there any bad reason for doing good if, in fact, Christian action is purely done for end results? In point of inalienable fact, doing good is its own reward, as many of us have learned.

---

[1] BU

The recipients simply and properly say, "Thank you!" I have never heard anyone ask, "Why did you help me? Why were you kind and loving? What was your motivation?" Have you?

> It is He who sits above the circle of the earth, and its inhabitants are like grasshoppers, who stretches out the heavens like a curtain, and spreads them out like a tent to dwell in. (Isaiah 40:22)

The book of Isaiah was written between 740 and 680 BC,[1] hundreds of years before the Greeks postulated that the earth was round.[2]

> Before I formed you in the womb, I knew you. (Jeremiah 1:5)

Physicians, scientists all, rely on the human heart as their first indication of a living person. A stethoscope is the primary tool of a general practitioner. In their mothers' wombs, babies' hearts begin to beat after only six weeks of development and growth. Every baby has a unique set of chromosomes and DNA that define a person.[3] How can anyone say that is not a person? This is science perverted by lies, beginning with Norma McCorvey who perjured herself in court, claiming that she was raped in Roe v. Wade. She was not, and Norma worked diligently until her death to undo the wrong she initiated under pressure from *feminists*. The Supreme Court decision should be overturned on the basis of false testimony. It was in fact reversed in 2022.[4] There are men who are in prison for murdering their unborn babies. Women who have their babies murdered are martyrs to the godless left, the counterpart to the *religious right*.

---

[1] CRS Reports
[2] World History
[3] Carlson
[4] CNN

> And Jesus answered and said unto them,
> "Take heed that no man deceive you." (Matthew
> 24:4)

Wisdom excelleth folly as far as light excelleth darkness. The Creator of mankind does not want anyone to be fooled; he does not wish that we be gullible or duped by clever sounding words, whether they be from scientists wearing white robes or priests wearing white robes. It is the essence of science to search for truth, ask questions, and seek to answer them. Toward this end, we must acknowledge the extremely transitional nature of scientific *fact* and the pretexts of contemporary scientific infallibility. Nowhere is the attempt to deceive us more malicious and destructive than when it seeks to deny the very existence of our Creator and separate us from him permanently.

> Verily, verily, I say unto thee, We speak that
> we do know, and testify that we have seen; and
> ye receive not our witness. If I have told you
> earthly things, and ye believe not, how shall ye
> believe, if I tell you of heavenly things? (John
> 3:11, 12)

The timeless veracity and wisdom of this statement is prophetic. After over two thousand years, what is often called the *myth*, the Holy Bible, has not been discredited. To the contrary, the obvious and manifest benefits to mankind of Christianity have been repeatedly addressed by professionals in psychology, criminology, and other modern disciplines. The deep and abiding science, beginning with the first sentence in the first book, is compelling testimony. For these reasons and more, Christianity is today the most popular human organization there has ever been. Yet many still *receive not our witness*. How can this be? Why will mankind not learn?

> If you were to take...all the authoritative
> articles ever written...on the subject of mental
> hygiene, if you were to combine them and refine
> them and cleave out the excess verbiage, if you

were to take the whole of the meat and none of the parsley, and if you were to have these unadulterated bits of pure scientific knowledge concisely expressed by the most capable of living poets, you would have an awkward and incomplete summation of the Sermon on the Mount. (*A Few Buttons Missing: The Case Book of a Psychiatrist*, by James T. Fisher, MD)

I say unequivocally that the evidence for the resurrection of Jesus Christ is so overwhelming that it compels acceptance by proof which leaves absolutely no room for doubt! (Sir Lionel Luckhoo, listed in the *Guinness Book of Records* as the most successful attorney in history)[1]

There remains, therefore, no supposition possible to explain the recorded phenomenon (of Christ's crucifixion) except the combination of the fructification and rupture of the heart. Samuel Houghton, MD, a great physiologist from the University of Dublin.[2]

---

[1] Conservapedia
[2] McDowell

# CHAPTER 7

# Hold Fast That Which Is Good

Test all things; hold fast that which is good.
(1 Thessalonians 5:21)

Is there a more concise explanation of the scientific method? *Test all things?* The essence of modern science is the proof of testable hypotheses, as scientists like to say. Such technique was proposed in the scriptures, which are mocked by so many who are proud of their lack of belief.

For every house is builded by some man;
but he that built all things is God. (Hebrews 3:4)

This is confirmation of Genesis 1:1. *Natural Theology*, written in 1809 by William Paley, makes a similar argument from design. Just as a Boeing 747 can never be the product of an explosion in a junkyard[1] neither can the far more intricate construction of a human being be the ultimate product of random mutations. Atheists pretend that "selection" confers creative powers of design on random mutations, but decades of empirical science prove otherwise.[2] (See also www.2001Principle.net.)

For ye have the poor with you always, and
whensoever ye will ye may do them good: but
me ye have not always. (Mark 14:7)

Although not strictly *scientific*, the foregoing verse demonstrates an important economic truth notwithstanding the astonishing achievements of modern science. Had one the vision to see forward

---

[1] Libquotes
[2] Linton, A. H.

more than two thousand years to witness the abundance of food and material goods we now produce, surely one would have surmised that *the poor* would be no more than a relic of the ancient past. Not so. In fact, nowhere are *the poor* more common and more destitute than in socialist countries led by atheist despots.

> (God) hath made of one blood all nations of men for to dwell on all the face of the earth, and hath determined the times before appointed, and the bounds of their habitation. (Acts 17:26)

Notwithstanding the vast differences in outward appearance of the peoples of the earth—from fair-skinned Norwegians to African pygmies—*all nations of men* are Homo sapiens. All humans can, within the constraints of blood typing, receive transfusions of the blood from men of *all nations*. This close biological relationship of mankind was cited before blood types and biological classifications were ever imagined. Moreover, Charles Darwin repeatedly expressed his disdain for what he considered to be the *savage races*, such as Africans.

"At some future period, not very distant as measured by centuries, the civilized races of man will almost certainly exterminate and replace, the savage races throughout the world" (*The Descent of Man*, Charles Darwin, chapter 3). This vile racist, who was an admittedly mediocre student in school, originated the pseudoscientific hypothesis of common descent, now called *evolution* or *Darwinian evolution*.

Today, Darwinian evolution is endlessly repeated to be *fact* by atheists.

> The irony is devastating. The main purpose of Darwinism was to drive every last trace of an incredible God from biology. But the theory replaces God with an even more incredible deity—omnipotent chance. (Theodore Roszak, *Unfinished Animal* (1975), pp. 101–102)

> For you (Israelites) are a people holy to the
> Lord your God. Out of all the peoples on the
> face of the earth, the Lord has chosen you to be
> his treasured possession. (Deuteronomy 4:2)

Today Israel prospers and shares its wisdom and goodness with the world. Twenty percent of Nobel laureates, notably in the sciences, have been of the Jewish faith, from a population representing only .2 percent of the world. No coincidence here, but this is divine providence foretold two thousand years ago. If Israelites are indeed God's chosen people, why then did he permit the Holocaust? This question mirrors the broader conundrum, "Why does God permit evil?" I have no answer beyond this.

> For my thoughts are not your thoughts,
> and your ways are not my ways, saith the Lord.
> (Isaiah 55:8)

> Every one that is of the truth heareth my
> voice. Pilate saith unto him, What is truth?
> (John 18:37–38)

Eighty-two percent of Americans believe that Jesus Christ was God, according to a survey taken by Princeton Survey Research Associates on December 2 and 3, 2004, for *Newsweek Magazine*. Jesus Christ said that people of the truth hear his voice. Today one especially hears atheists reply just as Pontius Pilate did two thousand years ago with a mockingly dishonest question: "What is truth?" when even a ten-year-old has the capacity to know what truth is. Some things never change.

Homosexuals across America have recently stepped up their evil efforts to deny the science of our birth. Men can become women and vice versa; they lie so reprehensibly to the devastation of countless thousands of young lives. Supreme Court nominee, Ketanji Brown Jackson, was asked in confirmation hearings, "Can you provide a definition of the word *woman?*"

Her incredible response was, "I can't. I'm not a biologist."[1]

---

[1] Daily Wire

Thus is truth turned on its head in these times of incredible scientific advance and technical sophistication.

# CHAPTER 8

# His Word Is Eternal

> Heaven and earth shall pass away: but my words shall not pass away. (Luke 21:33; Matthew 24:35)

Just as Genesis 1:1 foretold the scientific truth of the big bang or as it is called today, the Moment of Creation, so, too, does the Gospel of Luke describe the passing away of matter. Stars like our own sun are consuming their own hydrogen, fusing it into helium, losing vast amounts of mass, and radiating that energy into the heavens. All starlight will eventually pass away, and the universe will ultimately become dark and absolutely cold (*heat death*), devoid of life as we know it. Numerous massive black holes have been indirectly observed. They attract matter and light into their domain, never to be seen again. Or so scientists thought before they once again changed their mind. It is now thought that black holes can radiate away energy, which is to say, mass. Even black holes seem to pass away.

The duration of primary elements of matter we call protons has a lifetime estimated to be $10^{33}$ years.[1] The heaven and the earth shall pass into cold, dark oblivion, just as stated in the Holy Bible, a book ceaselessly mocked and ridiculed by the left. Thousands of years after the Bible said heaven and earth shall pass away, the second law of thermodynamics simply repeated it, in the name of *science*.

Universal heat death is the ultimate and final increase in entropy or disorder. The origin of our universe and its demise were both described in an ancient book with astonishing precision. The universe has not always existed, as previously thought by scientists, nor was it assembled

---

[1] Wikipedia

piecemeal. It was *created* with a big bang. The universe will not continue on forever; it will, it must pass away. These were not lucky guesses. They were inspirations from Nature's God. He made science. Our Founding Fathers had no doubt of that.

> Over half a century ago, while I was still a child, I recall hearing a number of older people offer the following explanation for the great disasters that had befallen Russia: "Men have forgotten God; that's why all this has happened." (Alexandr Solzhenitsyn)[1]

> I ask not for a lighter load, but for broader shoulders. —Jewish Proverb

[1] Templeton

# Endnotes

Endnote 1: By *the insuperable statistics of abiogenesis*, I mean the utter impossibility of any naturalistic explanation for synthesizing even a single polypeptide, much less the enormous variety of proteins, enzymes, and related structures which constitute every living cell. Our polypeptides (proteins) are so complex that even with a blueprint, contemporary scientists cannot produce them in a laboratory from reagents. How much more hopeless it would be to expect a pool of dirty water to produce hundreds of proteins necessary for the first living cell.

Humans are composed of over twenty thousand different proteins,[1] the largest of which is titin. It is 38,138 amino acid residues (sections) in a precise sequence.[2] The probability of assembling this sequence is 1/20 (amino acids) to the 38,138th power, times 1/2 to the 38,138th power, times 1/2 to the 38,138th power. There are D and L forms of amino acids, and we are made of only the L form; hence, the middle factor. Only peptide bonds are used in proteins, and they have a 50/50 probability because there are also nonpeptide bonds possible. This exponentiation brings the original naturalistic synthesis to about 1 chance in 10 to the 72,578th power. One chance in $10^{50}$ has been defined as *impossible* by prominent statistician, Émile Borel.[3] Billions and billions pale in significance to $10^{74,578}$, and remember, there are *only* about $10^{80}$ particles in the entire universe.[2]

Endnote 2: By *insuperable statistics of the Anthropic principle*, I refer to the several dozen physical constants, such as the gravitational constant, the strong nuclear force, and the fine structure constant, which are so finely tuned that if they were as little as one part in $10^{40}$ larger or smaller, there would be no planet earth and no human beings.

---

[1] NCBI
[2] OMIN
[3] Owlcation
[4] The New Scientist

The gravitational constant in particular is precise within one part in 10 to the $10^{120}$. What has been so desperately offered as an explanation for the Anthropic principle is the preposterous fantasy of *multiverse*, an infinite number of universes, of which we live in *just the right one*. It is not remotely connected to science or reality.

Profound Fortuitous Interdependencies, a term I originated in my book *Brilliant Creations: The Wonder of Nature and Life*, refers to the interrelated factors surrounding us that are utterly essential to life as civilized people know and enjoy. There is a pervasive elegance throughout our world and indeed our universe which far transcends the atheistic nihilism so widespread today in universities.

First example of Profound Fortuitous Interdependencies: Oxygen happens to be one of the more reactive elements, and it exists in the most reactive phase, viz. gas. This dynamic equilibrium producing our essential breath of life is possible only because

1. reactions are reversible;

2. the reversibility of such reactions is, so far as we can tell, 100 percent;

3. natural laws are consistent and harmonious and discoverable; and

4. plants recycle oxygen from water, not carbon dioxide. Carbon dioxide molecules are joined together to form carbohydrates, such as cellulose and sugar.

Second example of PFIs: The sophisticated engineering of matter, constructed primarily out of three simple building blocks, electrons, protons, and neutrons, is evidenced by the fact that only one part in $10^{17}$ is matter, and the rest is empty space. Moreover, these three foundational components can be combined with only slight variations to give us unique characteristics as found in metals, including high shear strength, malleability, ductility, high heat and electrical conductivity, abundance worldwide, high melting points, and other features without which there could be no airplanes, cars, or even internal combustion engines. As essential as metals[1] are to internal combustion engines,

---

[1] In contrast to water, plastic, glass, and fused quartz, which transmit light with near perfection, unlike metals or wood

[2] Mide

they would be virtually worthless without hydrocarbons to burn in them, which hydrocarbons just happen to be available in wondrous, underground storage tanks, to be refined and used worldwide.

But without oxygen and without its dispersal throughout our atmosphere, there could be no cars, no plane rides to faraway places, and no fires, even inside engines. Fire, a simple chemical reaction we have always taken for granted, is but one of the countless Profound Fortuitous Interdependencies.

Third example of PFIs: Vision made possible only through the combined PFIs of the elegant properties of electromagnetic radiation in the visible portion of that spectrum, including the capacity to transmit both energy as well as information, over unimaginable distances, with almost perfect fidelity and reliability, and the capacity to be bent by a wide variety of readily available transparent materials, so as to enable wide-angle vision instead of 1:1 tunnel vision of an image roughly the same size as the receiving retina, and all the while, photons pass through photons in every direction and in every wavelength, utterly unaffected in the slightest.

Endnote 3: This is the calculation for the cubic mile of air at sea level (standard temperature, which is, 0 Celsius): $(5{,}280 \text{ feet} \times 12 \text{ inches/foot})^2 \times (14.7 \text{ pounds/square inch} - 11.907 \text{ pounds per square inch at 1 mile})^2 = 11{,}212{,}469{,}453$ pounds or 5.606 million tons.

# Further Reading

## Books I Have Read and Recommend

*The New Evidence That Demands a Verdict* by Josh McDowell

*The Devil's Delusion* by David Berlinski

*The Irrational Atheist* by Vox Day

*Illogical Atheism* by Bo Jinn

*The Privileged Planet: How Our Place in the Cosmos Is Designed for Discovery* by Guillermo Gonzalez and Jay W. Richards

*Undeniable,* by Douglas Axe

*The Wonder of the World: A Journey from Modern Science to the Mind of God* by Roy Abraham Varghese

*The Case for a Creator* by Lee Strobel

*The Fingerprint of God* by Hugh Ross

*Brilliant Creations: The Wonder of Nature and Life* by John Phillip Jaeger

# Bibliography

AIImpacts. Accessed at https://aiimpacts.org/information-storagein-the-brain/.

AZ Quotes. Accessed at https://www.azquotes.com/author/3690Paul_Davies.

BigThinkEmpty. Accessed at https://bigthink.com/starts-with-abang/matter-mostly-empty-space/.

BigThinkBigBang. Accessed at https://bigthink.com/13-8/lemaitrepriest-proved-einstein-wrong/.

Borel, Emile. Accessed at https://quotefancy.com/emile-borel-quotes.

Brittanica. Accessed at https://www.britannica.com/science/sound-physics/The-ear-as-spectrum-analyzer.

BU. Accessed at https://www.bu.edu/files/2015/04/fundingbydiscipline550.jpg.

Carlson, Bruce M., *Patten's Foundations of Embryology*, 6th edition. New York: McGraw-Hill, 1996.

CNN. Accessed at https://www.cnn.com/2022/06/24/politics/dobbs-mississippi-supreme-court-abortion-roe-wade/index. html.

Conservapedia. Accessed at https://www.conservapedia.com/Lionel_Luckhoo.

CRSReports. Accessed at https://crsreports.congress.gov/product/pdf/R/R44283/14#.

Daily Wire. Accessed at https://www.dailywire.com/news/im-nota-biologist-supreme-court-nominee-says-she-cant-define-theword-woman.

EagerEyes.org. Accessed at https://eagereyes.org/blog/2015/the-island-of-knowledge-and-the-shoreline-of-wonder.

Edwards, Jonathan. Accessed at https://www.azquotes.com/ quote/1318369.

Etymonline. Accessed at https://www.etymonline.com/word/science.

FamousScientists. Accessed at https://www.famousscientists.org/ how-hertz-discovered-radio-waves/.

Fisher, James T., and Lowell S. Hawley. *A Few Buttons Missing: The Case Book of a Psychiatrist*. Philadelphia: J. B. Lippincott & Co., 1951.

Gizmodo. Accessed at https://gizmodo.com/georges-lemaitre-the-greatest-scientist-you-ve-never-h-1519769080#.

Goodreads. Accessed at https://www.goodreads.com/quotes/126931the-greatest-discovery-of-any-generation-is-that-human-beings.

Goodreads (Heisenberg). Accessed at https://www.goodreads.com/ quotes/9627737-the-first-gulp-from-the-glass-of-naturalsciences-will.

Hubble, Edwin. Accessed at https://www.nasa.gov/centers/marshall/ history/hubble02.html.

Lennox, John. "A Matter of Gravity." 2015. https://www.youtube. com/ watch?v=l63-fkyDtOc&t=102s.

Lewis, C. S. Accessed at https://www.azquotes.com/quote/811307.

Libquotes. Accessed at https://libquotes.com/fred-hoyle/quote/ lbv4a8p.

Linton, A. H. "Scant Search for the Maker," *Times Higher Education*, 2001. London.

LogicallyFallacious. Accessed at https://www.logicallyfallacious.com/ logicalfallacies/Argument-from-Incredulity.

Love Expands. Accessed at https://loveexpands.com/quotes/jonathan-edwards-539362/.

Mayo Clinic. Accessed at https://www.mayoclinic.org/healthy-lifestyle/ stress-management/in-depth/positive-thinking/art20043950.

McDowell, Josh. *The New Evidence That Demands a Verdict*, Nashville, Tennessee, Nelson Electronics & Publishing., 1999

*Merriam-Webster.* Accessed at https://www.merriam-webster.com/ dictionary/plausibly.

Mide. Accessed at https://www.mide.com/air-pressure-at-altitude-calculator.

NASA. Accessed at https://spaceplace.nasa.gov/big-bang/en/.

NBC. Accessed at https://www.nbcnews.com/id/wbna7139039.

NCBI. Accessed at https://www.ncbi.nlm.nih.gov/pmc/articles/ PMC4889822/.

NeuronResearch. Accessed at https://neuronresearch.net/vision/ files/ adaptation.htm.

NewsWithViews. Accessed at https://newswithviews.com/give-meone-generation-of-your-children-and-ill-transform-your-country/.

*Newsweek*, 2004. Accessed at https://www.newsweek.com/poll-christmas-miracle-123211.

NIH. Accessed at https://www.nia.nih.gov/news/optimism-linkedlongevity-and-well-being-two-recent-studies.

Omin. Accessed at https://www.omim.org/entry/188840\.

Owlcation. Accessed at https://owlcation.com/stem/ Borels-Law-of-Probability.

Parentotheca. Accessed at https://parentotheca.com/2021/09/27/ flourish-martin-seligman-summary/.

Penrose, Roger. *The Emperor's New Mind.* New York: Oxford Landmark Science, 2016.

Penzias, Arno and Robert Wilson. 1964. https://www.smithsonianmag. com/smithsonian-institution/how-scientists-confirmedbig-bang-theory-owe-it-all-to-a-pigeon-trap-180949741/.

Physics of the Universe. Accessed at https://www.physicsoftheuniverse. com/scientists_hubble.html.

Physics of the Universe. Accessed at https://www.physicsoftheuniverse. com/scientists_lemaitre.html#

QuoteFancy. Accessed at https://quotefancy.com/quote/781636/.

QuotesLyfe. Accessed at https://www.quoteslyfe.com/quote/Atheism-is-the-natural-and-inseparable-part-1138465.

Reasons. Accessed at https://reasons.org/explore/blogs/average-joes-corner/nothing-actually-something-to-ponder-from-lawrence-krauss

Richter, Curt. "Drowning Rats Psychology Experiment: Resilience and the Power of Hope." 1957. https://worldofwork. io/2019/07/drowning-rats-psychology-experiments/.

Rosza, Theodore. *Unfinished Animal.* 1975. ThriftBooks, Atlanta, Georgia.

RoughlyDaily. Accessed at https://roughlydaily.com/2021/01/29/.

Sagan, Carl. *Cosmos.* New York: Random House, 1980.

ScienceDaily. Accessed at https://www.sciencedaily.com/releases/2006/07/060726180933.htm#:~:text=The%20investigators%20calculate%20that%20the,100%20million%20bits%20per%20second.

Science History. Accessed at https://sciencehistory.org/education/scientific-biographies/john-dalton/

Shalev, Baruch. *100 Years of Nobel Prizes.* Ocala, Florida: Atlantic, 2003.

Templeton. Accessed at https://www.templetonprize.org/laureate-sub/solzhenitsyn-acceptance-speech/.

TheCrimson. Accessed at https://www.thecrimson.com/article/2020/1/28/lee-history-god-harvard/.

The New Scientist. Accessed at https://www.newscientist.com/lastword/mg24732961-500-how-much-stuff-is-there-in-the-universe/?

Tipler, Frank J. *The Physics of Immortality.* New York: Doubleday, 1994.

Uni of Oxford. Accessed at https://uni-of-oxford.custhelp.com/app/answers/detail/a_id/121/kw/motto.

UpperCervicalInstitute. Accessed at https://uppercervicalinstitute.com/wp-content/uploads/2020/05/Human-Genetic-Potential.MASTER.UPPER_.CERVICAL.pdf.

WayOfLife. Accessed at https://www.wayoflife.org/reports/karl_ marx.php.

WebMIT. Accessed at https://web.mit.edu/2.972/www/reports/ear/ear.html.

Wikipedia., https://en.wikipedia.org/wiki/Proton

World History. Accessed at https://www.worldhistory.org/Isaiah/; https://www.worldhistory.org/Eratosthenes/.

World of Work. Accessed at https://worldofwork.io/2019/07/drowning-rats-psychology-experiments/.

WTAMU. Accessed at https://www.wtamu.edu/~cbaird/sq/2014/02/07/what-is-the-shape-of-an-electron/.

# About the Author

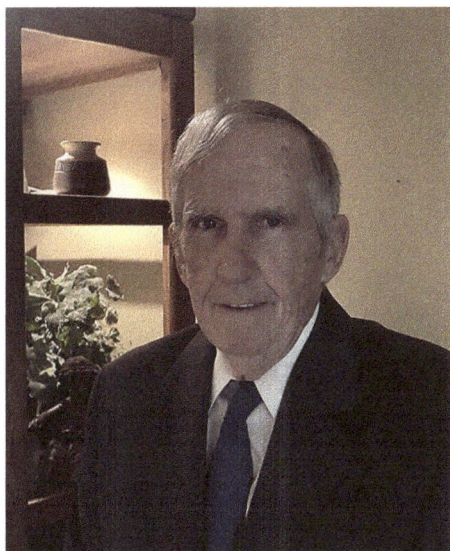

John Phillip Jaeger was born and raised in Granite City, Illinois, the son of a welder and a cook. Thanks to his brother, Gene Armes, who brought John to California to go to college, John earned a chemical engineering degree and a master's degree in business administration. He married his college sweetheart, and together they traveled the world with their two daughters. John is a certified diver, licensed pilot, triathlete, marathoner (PR 3:07), and expert snow skier who currently plays tennis six times a week using the Y-string racket he designed and built to minimize framing mishits (Non-Provisional Patent Application No. 17/706, 841).

He has been widely published in national newspapers and magazines to educate the public and correct misinformation, and toward that end, he finally published *Brilliant Creations: The Wonder of Nature and Life*, an inspirational science book described by Dr. John Orosz as *beyond incredible* and, by The Book Commentary, as *pure genius*.

www.ingramcontent.com/pod-product-compliance
Lightning Source LLC
Chambersburg PA
CBHW040910210326
41597CB00029B/5035